A Beginner's Guide To Understanding Technical Support

A Beginner's Guide To Understanding Technical Support

Jose-Albin D. Afable

Writers Club Press
San Jose New York Lincoln Shanghai

A Beginner's Guide To Understanding Technical Support

Writers Club Press
an imprint of iUniverse, Inc.

For information address:
iUniverse, Inc.
5220 S. 16th St., Suite 200
Lincoln, NE 68512
www.iuniverse.com

ISBN: 0-595-22574-8

Printed in the United States of America

To my wife Yvonne and
our children, Nicole, Gabrielle, and Justin.

Contents

List of Illustrations

List of Tables

Acknowledgements

This book would not have been possible if not for the support from very special people. These people have shaped my life; have given me the inspiration and determination to do my best to succeed.

I would like to acknowledge my parents (Marcelino and Norma), my wife (Yvonne) and our children (Nicole, Gabrielle, and Justin), my brothers (Abel, Albert, and Alex) and sister (Gerlie).

Thank you for your support.

Introduction

What is technical support?

Technical support is a service. It's about assisting users with problems and issues pertaining to technology, specifically regarding computers. It also has to do with providing acceptable, effective and efficient resolutions.

A Beginner's Guide To Understanding Technical Support is a unique guide designed for anyone wishing to venture into the field of technical support. This book will give detailed information about the different areas and people within a technical support organization. It will look at the levels of support within the support organization and how they function, interact, and manage their service requirements. Overall, it will provide a basic understanding of the processes, entities, and issues within a support organization so that anyone wishing to know more about the support organization functions or wanting to pursue a career in this area receives the information they need to make their informed decision.

Today, technical support continues to evolve and is becoming more complex. The main reason for this is the businesses need to utilize newer technologies. In the past, businesses generally competed locally and were more centralized. Due to enormous growth and fierce competition, companies have become decentralized with offices in various regions of the country and the world. Supporting these organizations are by no means a simple task. It requires a great deal of time and resources. That's why the use of advanced technology is required to manage user and customer issues effectively and efficiently.

As you continue to read this guide, you'll realize that a good portion concentrates on computer users. The users play an integral role in the technical support picture. They are the reason support organizations exist. Since they are the customers, support services are geared towards their needs. The support organization must make every effort to ensure their customers' satisfaction. This is why excellent customer service skills are required for support analysts.

Technologies continue to change the way we live and do business and the way we live our lives. Since technical support evolves with those technologies, it is important and clear that we, as technologists gain a thorough understanding of it in order to provide the best support possible. This book will inform you of the different areas within technical support, walk you through the process of how issues are resolved, and hopefully motivate you into creating new ideas to improve certain aspects of the industry.

CHAPTER 1

LEVELS OF SUPPORT

This chapter will provide in-depth information about the different levels of support within a technical support organization.

Many years ago technical support was simple because technology was limited and therefore support organizations only required a handful of technical people to manage it. The technologies used by businesses today require significant expertise and dedicated people to support them. Due to the complexities of business and technology, support organizations have evolved into complex entities as well. Today, it is essential that users and people in general grasp a basic understanding of technology. What better way to do it than to understand technical support? Understanding how a support organization functions can be difficult because of the many different entities and levels involved in the support process. Hence, it is important to understand what these different levels of support are and how they relate to one another. An understanding of each level is necessary because it forms the very foundation of the support organization. This foundation changes from one support organization to the next. It is important to remember that the number of support levels will vary depending on the size of the company. It is very likely that the larger the company, the greater the number of support levels. This is readily apparent when you compare a company with 100 employees to another company with 1,000 employees.

There are four distinct levels of support (**Table 1**). These support levels are Level 0 thru Level 3. Each level's responsibilities and functions vary. Trouble tickets which are electronic forms created to record the problem are escalated to each level depending on the type of technical problems or issues reported. These levels along with their functions are shown in Table 1.

Level	Function
Level 0	Self-Service Support
Level 1	Helpdesk
Level 2	Field Support
Level 3	Server Administration

Table 1 – Levels of Support

Level 0

Level 0 is the first level of support and is also known as self-service, e-support or online support. This level provides users with the ability to resolve issues on their own by using self-help related information such as online documentation, printed instructions, frequently asked questions (FAQs), and other documented support guidelines and resources. Companies with the resources available to create and maintain it are implementing this level. Level 0 support rarely involves person-to-person communication or interaction. Level 0 is the newest of the four levels of support. It was actually created just a few years ago with the advent of newer and faster technologies. With faster and more reliable technologies organizations were able to develop the tools needed to provide Level 0 support. One such example of a tool that has evolved through the use of technology is websites containing FAQs about the company's products, services, software and hardware information. Another example is an online tool which allows users the ability to request changes to system ids or to reset system passwords. As technology continues to improve, I predict that in a few years, these types of tools will drastically get better and evolve into something that will enable users to resolve many issues remotely on their own. The contents of these online tools vary from one company to the next because software technologies are not always 100% compatible. Therefore, it is difficult to implement this type of resource due to its complexity.

Please remember that these tools are not designed by any means as an attempt to diminish the level of support. It is simply a resource for users wishing to find answers or resolve issues on their own. These types of users, which will be covered in Chapter 2, have gained enough knowledge to resolve their own technical issues given the proper documented information. This is one reason why certain technical support organizations have implemented Level 0 as a way for users to get quick resolutions for their own technical issues. Self-Service Support alleviates the simple technical issues and allows more time to be devoted to projects that require special attention.

Level 1

For many support organizations, Level 1 forms the first line of support. Level 1 is also called the Helpdesk or Call Center. This level consists of direct phone support and involves person-to-person phone communication. Users normally call a specific phone number, which is directed to a support analyst. The analyst creates a trouble ticket for the user using a call tracking software and documents the user's personnel information such as name, location, phone number, and the issue reported. The analyst then proceeds to resolve the issue over the phone using developed online troubleshooting documents or his or her personal knowledge experience of the issue. When the issue is resolved, the analyst confirms the resolution, documents it on the trouble ticket, and closes the ticket. However, if the issue is not resolved and requires more advanced troubleshooting, the analyst escalates (see escalation process — **Figure 1**) the trouble ticket to Level 2 support with the aid of the call tracking software. The support level analyst receiving the escalated trouble ticket reviews the documented problem and proceeds to provide a resolution to the user. During this process the Level 2 analyst receiving the escalated trouble ticket will communicate with the user and the Level 1 analyst that escalated the trouble ticket. Everyone is informed to promote teamwork.

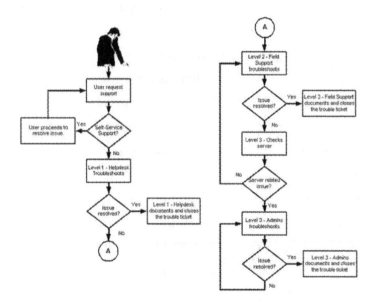

Figure 1 – Steps in the Escalation Process

Step 1 –   User requests support

Step 2 –   Can issue be resolved with Self-Service support? If yes, then the user resolves the problem. If not, proceed to step 3.

Step 3 –   User contacts Level 1 for support and the Helpdesk troubleshoots the problem.

Step 4 –   Did Level 1 resolve the issue? If yes, then Level 1 documents and closes the trouble ticket. If not, proceed to step 5.

Step 5 –   Issue is escalated and Level 2 troubleshoots the problem.

Step 6 –   Did Level 2 resolve the issue? If yes, then Level 2 documents and closes the trouble ticket. If not, proceed to step 7.

Step 7 –   Issue is escalated and Level 3 troubleshoots the problem relating to the server.

Step 8 –   Was issue server related? If no, issue is referred back to Level 2 (step 5). If yes, proceed to step 9.

Step 9 –   Level 3 continues troubleshooting.

Step 10 –   Did Level 3 resolve the issue? If yes, then Level 3 documents and closes the trouble ticket. If not, then proceed back to step 9.

Level 2

Level 2, also known as Field Support, handles a large portion of trouble tickets. In some larger support organizations, Level 2 is divided into three groups. These groups are Level 2 Field Support, Level 2 Call Center, and Level 2 Hardware Support. These groups have the task of providing face-to-face technical support. Level 2 support analysts must have at least 1 to 2 years of technical experience within their areas of expertise. They must be very knowledgeable of the products and services of the support organization. They must also have excellent customer service skills in order to communicate information to the user and extract the required input necessary to resolve the problems. Input from the user is needed to begin the troubleshooting process and necessary to gain an understanding of what the user was working on when the problem occurred. Knowing the details of what the user was working on prior to the issue is paramount to its resolution of that problem. To get these prior details, the support analyst must be able to properly communicate with the user. That is why excellent customer service skills are a necessity for any support analyst.

Normally, Level 2 resolves a high percentage of complex trouble tickets. These trouble tickets includes problems which Level 1 (the Helpdesk) could not solve and issues that Level 3 will not accept until Level 2 has attempted to resolve the problem first. Level 2 trouble tickets that are not successfully resolved are then escalated to Level 3 for resolution. It is imperative that level 2 trouble tickets are documented and that all available troubleshooting resources were used to resolve the issue prior to escalating it to Level 3.

Level 3
This level consists of a number of different groups:

- Server and database administrators
- Software developers
- Network infrastructure technicians
- Software testers.

These groups have highly specialized skills, which are only attained through extensive hands-on experience and education. They form the last line of support. Issues escalated to these groups become an immediate priority because it normally affects many users. Since these escalated issues are unresolved by the other support levels and probably were open issues for a while, they warrant immediate attention to further minimize the users' downtime. Common examples are those related to server and database access. Other issues which Level 3 handle include server administration, server backups, server restores, support for internally developed applications, and network connectivity. Level 3's main responsibility is server administration. That's what a bulk of their trouble tickets' category fall into. Since this level is specialized, the number of analysts is far less than what you'll see in the other levels. They handle fewer trouble tickets than the other levels.

By now you should be able to somewhat envision the structure of a technical support organization. And you should start to gain an understanding of some of the processes within your own support organization. While you may be able to see similarities within your own organization, it is an important note to remember that having all four levels of support is not always the norm. Some companies make specific distinctions (**Table 2**) between the different levels, while others combine all levels into one distinct group. No matter how many levels there are and how they are grouped, each analyst will have specific areas of responsibilities as well as experience.

Level	Function
Level 0	Self-Service
Level 1	Helpdesk or Call Center
Level 2	Field Support
Level 2	Call Center (Advanced)
Level 2	Hardware Support
Level 3	Server Administration
Level 3	Software/Application Development
Level 3	Network Infrastructure

Table 2 – Support Levels for Larger Companies

Call Tracking Software

There are various call tracking software that is available for support organizations to monitor trouble tickets. Some international companies have even implemented global call tracking systems. These advanced systems allow management the ability to perform various analyses to make management decisions such as increasing technical support services in specific regions. The call tracking software can also provide the following features and advantages:

- Measure the length of time it takes to resolve an issue
- Quantify the number of issues resolved within a specific time period
- Provide historical records regarding each user and the type of issues reported
- Creates a knowledgebase that analyst can use to resolve recurring issues
- Provide a breakdown of which products or services are being utilized

Call tracking software is designed specifically for service and support organizations. There are features built within the software, which allow organizations to create inboxes for each support analyst or each level of support. These inboxes are used for trouble ticket escalation. When I mentioned escalation earlier, I was merely referring to a documented trouble ticket being assigned to an inbox. Each support analyst and each level of support maintain their own inbox. When a trouble ticket has been escalated, it is collected within a support level's inbox. Trouble tickets that are escalated to a support level inbox are retrieved by a support analyst and re-assigned to their own personal inbox. An inbox is basically a repository where trouble tickets are collected until an analyst retrieves it and a resolution is provided.

Once a trouble ticket has been escalated to a higher level of support, it is not by any means final. The higher level can reassign the trouble ticket back to the lower level if it discovers that the issue does not warrant higher-level assistance. For example, if Level 2 escalates a server related issue to Level 3 and the server administrator checked the server and found it to be in working order, the server administrator can send the trouble ticket back to Level 2 for further investigation or troubleshooting. Both groups will work together until the problem has been resolved.

Understanding these four basic levels of support is essential to understanding technical support. It is the basic foundation of any support organization. This foundation along with other elements is what makes a strong and efficient support organization. Without them, the foundation is weak and the processes become inefficient and ineffective.

CHAPTER 2

USERS

This chapter focuses on the different types of users and how to deal with them appropriately.

2.1 Types Of Users

When I first started in this industry as a technical support analyst, I was unaware of any distinctions between users and their personalities. It wasn't until later that I realized that there are different types of users. I was aware of there being different types of customers. But I did not equate this into technical support. So as I continued, it was not until later in my years in technical support that I realized the importance of understanding users. It wasn't just about resolving their problems. It's about providing the best support service you can possibly provide. I then realized that each of them is unique in their own way and getting to know them helps in the long run.

Knowing the type of user is relevant to the successful resolution of a reported technical problem. Users are customers and a required element of the technical support organization. They are the buyers of support services. That is why excellent customer service is a required skill for any technical support analyst to resolve reported problems. An analyst must be able to interact and communicate effectively with a user and must also be able to effectively manage that user's expectation. These skills are important because lack of them impedes on the satisfactory resolution of the reported issue and makes the education, training, and experience of an analyst useless.

In the next few paragraphs, the following types of users will be defined:

- Novice
- Intermediate
- Risky

- VIP
- Busy

There are some users, who fall into a combination of these user types, while others will fit strictly into one particular type. This is an important factor when you start to formulate an approach to dealing with users. You'll understand as we go along the importance of knowing and understanding users.

Novice

The first user type is the novice. The novice is either a user who has very little knowledge of computers or a new employee who hasn't had the opportunity to learn technologies available internally. Therefore, this user often requires special assistance. At times novice users are easily stressed when an issue arises and are very reluctant to change and embrace new technology. This is understandable given the pressure they are in to quickly learn their job. The support organization should provide training in order to avoid many of the problems that novice users encounter. However, there will still be issues that arise from the user's fear of technology. Some users have had limited or no exposure to technology and are reluctant to learn it. Therefore, it is important for analysts to recognize the users' reluctance and provide the necessary training.

Intermediate

Intermediate users have a keen understanding of technology and are able to resolve minor issues personally. They have many years of experience working on computers and other peripherals. They are not a novice and have probably worked on computers in their prior jobs. Unlike novice users, intermediate users that are new to a company lack the understanding of business processes not technology. Given the time to get acclimated, they can be productive very quickly. They are frequent users of Level 0 support. The intermediate user will seek assistance from the Helpdesk only if the issue is an emergency or they've run out of resources to resolve their problem.

Risky

Throughout our lives we've had to deal with difficult people. That same thing applies in technical support. There are difficult users. These users are what I call "risky". Probably the more problematic type of user within this group, the risky user, has attained a substantial knowledge of technology and perceives themselves as "experts". They can resolve minor and sometimes complex issues themselves. The problem, however, is they will sometimes go too far and beyond their expertise causing problems to multiply. In my experience, I've encountered risky users accessing restricted areas that cause damage to the computer's operating system.

I am not trying to paint an unpleasant picture of risky users. I am merely making you aware that they exist and knowing of their existence will give you a better understanding of how to assist them appropriately. Risky users from my experience don't always pose a big problem. Just letting risky users know what they can and cannot do can minimize problems dramatically.

VIP

The fourth type of user is the VIP. The VIP is a user that is normally in upper management. Due to their level within a company, their problems require immediate assistance. They embrace new technology more easily than other users because their busy schedule warrants the use of technology. If technology can alleviate some of the manual processes that they do, they welcome it, because saving time allows them to concentrate on other important matters. VIPs are generally easy to support. When assisting them you'll find that they want to know the cause and resolution of their issues so they can easily fix the problem if it occurs again in the future. Even with their enormous responsibilities, VIPs will make an effort to assist the technical support analyst in resolving their technical issues.

Busy

The last type is the "busy" user. The busy user is very pre-occupied. This user requests assistance with an issue but is too preoccupied to help the analyst resolve it. In some cases, the busy user forgets the reported issue and only remembers it when the matter becomes urgent. Often the analyst makes several trips to the user's location until the issue is resolved. If numerous attempts were made to visit the user and the user fails to allocate the time needed, the analyst may close the trouble ticket. In closing the trouble ticket, detailed documentation must be entered noting the failure of the user to respond to the repeated request. A message should also be sent to the user with information on why the trouble ticket is being closed. Closing the trouble ticket after three visits should be the last option and will sometimes force the busy user to acknowledge the request.

Knowing the types of users the technical support analysts deal with can better prepare them when assisting the users. Some analysts must be well prepared to resolve the issues reported regardless of the types of users reporting the issues.

2.2 Dealing With Users

Through all my years of working in Information Technology, I've discovered technical support to be the most difficult and demanding. The reason is because technical support analysts are constantly relegated to resolving one computer problem after another. Day in and day out they arrive to work facing service complaints and computer problems. To some technical support analysts, this can be an unbelievable burden. That's why technical support requires unique analysts that can manage time and handle the stress associated with their positions.

Time and stress management are important issues that support analysts face daily. It's important that they handle it accordingly. Part of managing time and handling stress is to deal with users effectively and tactfully. Technical support analysts have to be tactful when dealing with any user. The analyst must be able to approach a user with a clear mind and good intent. They need to be able to focus on the problem at hand. If they don't, the problem will severely get worse and cause more stress on themselves and the user.

To relieve stress a support analyst must deal with the user appropriately. How do support analysts deal with the user? The first thing an analyst must do is to try identifying what type the user is. This can be accomplished by doing research such as looking up past trouble tickets and finding out what types of problems, if any, the user reported previously; checking personnel directory listings to find out the user's personnel classification; check with other analysts that have assisted the user before. The support analysts' goal is to find out what type of user the person is so that they can formulate a strategy on how to deal with the user.

As we discussed earlier, there are five types of users. In the next few paragraphs we'll discuss steps on how an analyst deal with each types of users to help himself or herself provide the best service possible. Please remember that in certain situations an analyst may assist users, which exhibit different types of behavior and fall into a combination of user types. In this case an analyst will need to formulate a combination strategy of dealing with those user types.

Dealing with a novice requires time and dedication. That's because a novice is generally someone with very little computer knowledge or someone that is new to the organization. In this case dealing with this type of user means making them familiar with the internal policies and procedures along with service expectations. To be effective in dealing with a novice, analysts must make sure that they speak clearly so that the novice can properly understand what the analyst is trying to relay to them. An analyst also needs to provide the novice with a thorough explanation of the problem. Because in some instances, the novice doesn't even know what their problem is and how they caused it. By doing this, the novice becomes confident in the analyst's expertise and allows the novice the opportunity to participate in the resolution of the problem.

If an analyst feels that during the process, the novice is not able to comprehend what he or she is talking about and the analyst has spent substantial time thoroughly explaining it, then the analyst should suggest training for the user. The analyst should recommend this tactfully and not belittle the user. That's the last thing an analyst would want to do because doing so will only cause further headaches and stress. An analyst should take the time to earn the novice's confidence. With time and proper guidance, novice users can learn enough knowledge to resolve problems on their own and become an intermediate user.

Since an intermediate user only requires assistance when needed, they will only call when they've encountered problems they've never had before. The intermediate user will be able to give an analyst some information on how the problem occurred. They'll be able to provide more details and give the analyst the time he or she needs to resolve the problem. It is important in this case that the analyst takes the time to resolve the problem because it is generally a difficult one. In doing so, an analyst needs to provide a clear explanation of what caused the problem and how to prevent it in the future. In this case, after the problem has been resolve, the analyst should forward any pertinent documentation to the user pertaining to the resolution of the problem so that the user can resolve the problem in the future without calling technical support for assistance. If the analyst's resolution is too complex for the user to understand, the analyst should inform the user to contact the helpdesk if the problem reoccurs. This complex resolution should then be documented in the trouble ticket so that the issue can easily be resolved next time around.

When dealing with the risky user, an analyst should schedule enough time for the resolution process. That's because the problem will require a great deal of time and resources on the analyst's part. Prior to visiting the user, the analyst should make sure that he or she research the issue beforehand to find out what problems, if any, the user has reported in the past. By doing this, the analyst yields vital information about the user, which he or she can use in formulating a solution strategy. The analyst might discover that the user has called about the same problem more than once before and probably have been provided a resolution already. When assisting the risky user the analyst should be focused and not seem upset. Being upset will only make the solution process difficult. Instead the analyst should maintain the formulated strategy and try to get as much detailed information from the user about the problem. This is very important and will weigh heavily on how quickly it will take to resolve the problem. Because this type of user has a knack for creating difficult issues, they might not tell the analyst how the problem occurred. If this is the case the analyst should be tactful and try to gain the risky user's confidence. In doing so the analyst might be surprised and have the user admit their fault and give the analyst the vital information he or she needs to resolve the problem.

In my experience, I've encountered many instances when a risky user continually caused the same problem over and over again. How did I deal with this situation? After continued warnings, I gathered all of the pertinent documentation on the issues and met with his supervisor. I informed his supervisor about his continued problem and interjected that we cannot continue to spend any more of the company's valuable time and effort on problems he continues to commit. The outcome was, he stopped causing problems. Please keep in mind that this should be the last resort and should only be done when there are complete and detailed documentation on the issue along with prior management approval.

The VIP user has to be dealt with carefully. That's because they are normally in top-level management within a company. Most technical support analysts are very reluctant to support VIP users. In some cases, it is because they know that if they make a mistake, the VIP has the authority to make the job difficult for the analyst. However, it is my view that if you please the VIP user, he or she can be a powerful ally. A VIP user can provide you with positive feedback of your service, which you can use in your evaluation.

VIP users are actually not difficult to support. They are very knowledgeable. However, an analyst must be prepared when assisting the VIP user. The analyst must gather enough information to resolve the issue prior to meeting with the user and must update the user about all aspects of the resolution process. The analyst also needs to be confident and an expert in technical support because VIP users will not hesitate to let an analyst know that he or she is incompetent.

To assist the busy user, an analyst needs to be prepared to make a number of trips to the user's location. The analyst needs to ensure each trip to the user's workstation is documented and any events are logged in the trouble ticket. This information will provide a strong defense if for some reason it is needed. The analyst must also leave a message or note to the user for each trip that he or she made to the user's location. After three trips, the analyst should inform the user that he or she would close the trouble ticket and inform the user to contact the helpdesk again for further assistance. Doing this will normally make the user respond to the analyst's repeated attempts.

Regardless of which type of user an analyst supports, it is very important that the analyst follow up with each of them. This is part of great customer service and goes a long way in letting the users know that they are important. It is also good marketing to let the users know of other services that the support organization offers such as training, after-hours support, etc. Remember as a technical support analyst, you are the support organization's best salesperson of its products and services.

Dealing with users can be unpredictable. That's because they all have different personalities and demeanors. As we've discussed, knowing which type you'll be assisting will help you formulate a strategy of how to deal with them. This strategy is vital to winning the user's confidence in your ability and helps alleviate further problems in the future. The strategy that you must follow should cover all steps from knowing your user and conducting a follow-up after the problems has been resolved. Below is a 9-step process you can follow in resolving your user's problem.

Step 1 -   Know your user's type as described previously.

Step 2 -   Read and understand the user's reported problem prior to visiting the user and try to come up with a tentative solution, if possible.

Step 3 -   Be cordial when you arrive at the user's location.

Step 4 -   Reiterate the reported problem to clarify with the user so that the user does not have to describe the problem again.

Step 5 -   Ask the user what he or she was working on when the problem occurred.

Step 6 -   Begin troubleshooting.

Step 7 -   Keep user informed of the progress (If the time limit given by the user has expired, ask the user if it can be rescheduled or continue troubleshooting.)

Step 8 -   Once the problem has been resolved, explain what caused the problem, how the problem was resolved, and how to prevent it in the future.

Step 9 -   Follow up with the user to ensure that the resolution given is working.

As we've discussed, dealing with users is a full-time job and the majority of an analyst's daily responsibilities. This requires good technical and customer service skills. We've already covered some of the traits that an analyst should possess. However, there are still other traits that an analyst should possess.

These traits include:

- excellent organizational and prioritization skills
- professionalism
- ability to conduct training
- confidence to handle unfamiliar issues

In dealing with users an analyst must be able to organize and prioritize. As an analyst, there are enormous responsibilities, information, and communication that need to be addressed. Without the ability to organize and prioritize, the analyst will become lost and unable to provide the support services properly.

Providing services is the main product of the support organization that is usually performed in a corporate environment. It is important that all services are provided in a professional manner.

Troubleshooting is the process of correcting and finding resolutions to a problem. During troubleshooting, the analyst must explain to the user how the problem occurred and how to prevent it in the future. In doing this, the analyst is training the user. The ability to conduct training is essential in order to help the user understand the problem and the solution.

There are many different types of issues an analyst will encounter throughout his or her career. The knowledge and experience gained by analysts are a direct result of resolving issues. In order to gain additional experience, an analyst must be able and not be afraid to handle unfamiliar issues. Handling complex and unfamiliar issues is one of the best ways to gain additional knowledge.

Having these traits and mastering them will afford the analyst the opportunity to excel. Possessing these traits, an analyst will have a good foundation to succeed in his or her career.

CHAPTER 3

MANAGING USER EXPECTATIONS

This chapter will cover managing and measuring user expectations.

3.1 Service Level Agreements

To completely understand technical support you also need to be aware that there are other elements that are part of a support organization. One such element is user expectations. There are important things that are expected by users. It's not only about trying to provide the best technical service. It has to do with first knowing what providing the "best service" means. In order to know this, concrete measurements are needed by the support organization.

These measurements are gathered by getting feedback from the user community on what their expectations are. The feedback and expectations are then combined in an agreement called the service level agreement (SLA).

The SLA contains the following information regarding the technical support services to be provided. The SLA contains information such as:

- Service period
- Estimated number of trouble tickets to be resolved
- After-hours/non-business hours support guidelines
- Time frame for trouble ticket resolution
- Priorities of trouble tickets
- Cost of support services.

The SLA will state the service period or the length of time the agreement is to remain in effect. Normally, the agreement is discussed and renewed every fiscal year. Other times, it's permanent and is reviewed only when substantial changes are needed.

The number of trouble tickets is used to determine the cost of services. The support organization reviews the number of trouble tickets from prior years to determine whether to increase or decrease the cost for services.

Many large companies have departments working around the clock. In order to service these departments, support organizations must have analysts available to provide technical support. In many circumstances, it is not feasible to have dedicated analysts to service a limited number of users. This is where after-hours support comes in. The SLA has provisions on how after-hours support is provided.

The SLA also contains the time frame for trouble ticket resolution. It gives guidelines on how priorities are assigned and how long it will take to resolve the issues, based on their priorities.

Another information that is contained in the SLA, which is very important, is the cost of services. A detailed description of services and their costs are outlined in the SLA.

The SLA is a contract between the support organization and the user community. It becomes the guidelines by which all support organizations must abide by to provide the best technical service possible. A support organization is unable to measure how their products and services are being delivered without the SLA.

3.2 Measuring Support Services

Customer feedback is another important element of technical support. It is the by-product of the service provided. Feedback comes in the form of returned surveys. It's an important performance measure for the support organization, because it is a direct response from the buyers of the services, the users. In almost any organization, measurements are done with the aid of surveys.

Surveys are conducted in many ways. A sample survey is shown in **Figure 2**.

- Electronic forms sent via email

- Hard copy or printed forms sent by interoffice mail
- Regular email communications to users
- Phone conversations with users
- Web based forms

The survey should include pertinent information regarding the support service that was provided. It should include information contained in the trouble ticket such as:

- Trouble ticket number
- Name of the support analyst that provided assistance
- Detailed description of the reported problem
- Name of the user
- User's phone number
- Date the problem was reported
- Date the problem was resolved

The survey questions should measure the timeliness of the response to the problem and the attentiveness of the support analyst resolving the problem. It should contain rated questions or topics such as:

- Timeliness of the response

- Timeliness of the resolution

- Understanding of the problem by the support analyst

- Professionalism of the support analyst

- Satisfaction in the resolution

- Overall quality of the service provided

Ticket #:		User Name:	
Support Analyst:		Phone:	
Reported Problem:		Date:	

In order to provide the best service to our users, the Support Organization is interested in your feedback for the service you recently received. Please take a few minutes to complete this survey. Thank You.

Please use the following scale to rate the service:
4 – Excellent
3 – Very Good
2 – Good
1 – Unsatisfactory

Please rate the support analyst based on the following topics.

1. **Responded to your problem in a timely manner**
   ☐ 4 – Excellent
   ☐ 3 – Very Good
   ☐ 2 – Good
   ☐ 1 – Unsatisfactory

2. **Resolved your problem in a timely manner**
   ☐ 4 – Excellent
   ☐ 3 – Very Good
   ☐ 2 – Good
   ☐ 1 – Unsatisfactory

3. **Was pleasant to work with**
   ☐ 4 – Excellent
   ☐ 3 – Very Good
   ☐ 2 – Good
   ☐ 1 – Unsatisfactory

Figure 2 – Sample Survey

Surveys are excellent tools designed to compile data for measuring customer satisfaction. Surveys are also necessary to gauge the organizations' compliance with service level agreements. They provide a measure for management to ensure that the organization is meeting its SLA requirements. They also give analysts a status of how their individual services are received by each user that they provided the service to. Another important aspect of surveys is that it allows organizations to identify and segment groups of users that may require special support services and to correct problem areas, which support analysts, may not be aware of.

There are helpful measurements surveys provide. One aspect of surveys, which cannot be measured, is the sense of importance users receive when they realize that their feedback is valued and accepted.

CHAPTER 4

SUPPORT LEVEL RELATIONSHIPS

This chapter provides insights into the technical support organization and covers relationships and responsibilities.

4.1 Managing Support Level Relationships

The support organization must work together as a team in order to provide the best service to users. Each level of support must adhere to the SLA and ensure that services are in accordance with SLA guidelines.

Communication is key to a support-driven organization. Each levels of support must be able to communicate effectively, especially when assisting thousands of users.

Although support organizations are divided into different levels, they must work as one cohesive unit to accomplish their daily responsibilities. The organization must be able to efficiently manage issues that are reported by users. Having an understanding of each other's roles and responsibilities and how they interact with one another is critical to the success of the organization and its services.

Part of understanding technical support is knowing how each support levels interact with one another and the roles each level plays. Each analyst within each support level must know their own responsibility as well as their group's area of responsibility. They should know their boundaries and differentiate where the other group's responsibilities begin and ends. Knowing each support level of responsibility (**Table 3**) will enable each analyst to properly escalate any issues reported by users.

SUPPORT LEVEL	RESPONSIBILITIES
Level 0 Self-Service	Online documentation, content and all services related to self-service
Level 1 Helpdesk	Hardware and software phone troubleshooting support
Level 2 Field Support	Advanced hardware and software field troubleshooting support
Level 2 Call Center	Advanced hardware and software phone troubleshooting support, and internal applications
Level 2 Hardware Support	All hardware related issues (desktops, laptops, servers, printers, etc.)
Level 3 Server Admin	Servers, network backups and restores, network ID setup, and network printer setup
Level 3 Development	Development and support of internal applications and programs
Level 3 Infrastructure	Network connectivity, routers, switches, and data wiring

Table 3 – Areas of Responsibilities

Interactions between each level are an important daily occurrence within the support organization. Regularly, each level must interact with one another regarding issues reported by users or other support levels.

Knowing whom each level interacts with is essential to understanding technical support relationships.

For example, let's look at Level 0 Self-Service. This support level will generally interact with all levels of support regarding the Self-Service support content such as frequently asked questions, troubleshooting steps, etc. Level 1 Helpdesk on the other hand will generally communicate only with Level 2 Field Support, Level 2 Call Center, and Level 2 Hardware. As you look at the list below, you'll get a sense of the interactions between each support level and what the interactions are based on.

Level 0 Self-Service interacts with

- Level 1 Helpdesk regarding self-service content and information for users
- Level 2 Field Support regarding self-service content and information for users
- Level 2 Call Center regarding self-service content and information for users
- Level 2 Hardware regarding self-service content and information for users
- Level 3 Server Administration regarding self-service content and information for users
- Level 3 Development regarding self-service content and information for users
- Level 3 Infrastructure regarding self-service content and information for users

Level 1 Helpdesk interacts with

- Level 2 Field Support in relation to trouble ticket escalation, high priority issues

- Level 2 Call Center in relation to trouble ticket escalation, high priority issues

- Level 2 Hardware Support for hardware related problems

Level 2 Field Support interacts with

- Level 1 Helpdesk on relevant updates such as server problems

- Level 2 Call Center on updates such as server problems

- Level 2 Hardware Support on escalated hardware issues

- Level 3 Server Administration on escalated issues such as servers not responding

- Level 3 Development on application errors

- Level 3 Infrastructure on escalated slow network connectivity

Level 2 Call Center interacts with

- Level 1 Helpdesk about updates such as server problems
- Level 2 Field Support about updates such as server problems
- Level 2 Hardware Support about escalated hardware problems
- Level 3 Server Administration about escalated issues such as servers not responding
- Level 3 Development about application errors
- Level 3 Infrastructure about escalated slow network connectivity

Level 2 Hardware Support interacts with

- Level 1 Helpdesk about updates pertaining to hardware issues
- Level 2 Field Support about updates pertaining to hardware issues
- Level 2 Call Center about updates pertaining to hardware issues
- Level 3 Server Administration to assist with server hardware issues

Level 3 Server Admin interacts with

- Level 2 Field Support to report server problems and downtime
- Level 2 Call Center to report server problems and downtime
- Level 2 Hardware Support to report hardware problems with server
- Level 3 Infrastructure to report server problems and downtime
- Level 3 Development to report server problems and downtime

Level 3 Development interacts with

- Level 2 Field Support to report application problems or resolutions
- Level 2 Call Center to report application problems or resolutions
- Level 3 Server Administration to report application's server requirements

Level 3 Infrastructure interacts with

- Level 2 Field Support to report connectivity problems
- Level 2 Call Center to report connectivity problems

- Level 3 Server Administration to report connectivity problems

As you can clearly see, there are many elements that form a technical support organization. For the most part it involves the user. But in certain cases, before user issues can be resolved, each level of support must be able to work with one another to help resolve those issues. Managing relationships between each support level is important to the support organization. It has to be done effectively and efficiently to minimize the downtime experienced by users. It is also important for each level of support to realize that they themselves become customers within their own organization. Therefore, each level of support should be given the same level of courtesy as they would any customer or user.

4.2 Support Level Positions

The structure of the support organization varies from one company to another. The difference is mainly in size and number of personnel. Understanding technical support requires knowledge of the personnel within the support organization. The list below will provide some details of the personnel within the support organization.

Level 0

- Content Manager
- Content Specialists
- Web Developers

Level 1

- Helpdesk Manager
- Helpdesk Team Leader
- Helpdesk Analysts

Level 2

- Desktop Support Manager
- Support Team Leader
- Desktop Support Analysts
- Hardware Support Analysts

Level 3

- Administration Manager
- Server Team Leader
- Server Analysts

- Software Development Manager
- Developer Team Leader
- Software Developers

All managers within each level reports to the Chief Technology Officer (CTO). This personnel detail is for informational purposes only and it is not meant to apply to a specific company. Each company is different and will have different personnel requirements. This information is an example to provide insight into personnel requirements within the support organization.

4.3 Support Level Personnel Requirements

The experience requirements of support analysts depend on the level they belong to. These requirements also vary from one company to another. To give you a better understanding, **Table 4** lists details of each personnel's minimum experience requirements.

Positions	Experience Requirements
Level 0	
Content Manager	3 years managing online contents. Development experience.
Content Specialist	Entry level to advanced experience with data and website information
Web Developer	Entry level to advanced software development experience
Level 1	
Helpdesk Support Manager	2 years of management experience, 10 years of overall technical experience
Helpdesk Support Team Leader	5 years in helpdesk support, 2 years in a lead role
Helpdesk Support Analyst	Entry level to 1 year of phone support
Level 2	
Desktop Support Manager	3 years of management experience, 10 years of overall technical experience
Desktop Support Team Leader	5 years of total desktop support, 2 years in a lead role
Desktop Support Analyst	Entry level to 2 years of desktop support
Hardware Support Analyst	2 years of hardware support
Level 3	
Server Manager	10 years of overall technical experience, 3 years management experience, must have server administration experience and server certifications
Server Team Leader	2 years as a server administrator with some lead responsibilities. Must have server certifications
Server Analyst	2 years as a server administrator. Must have server certifications
Development Manager	3 years managing development projects, 2 years development experience, certifications
Development Team Leader	2 years in lead development role. Must be certified
Developer	Entry level to 2 years development experience. Some certifications required

Table 4 – Personnel Experience Requirements

CHAPTER 5

THINKING ABOUT A TECHNICAL SUPPORT CAREER?

This chapter will discuss careers in technical support and provide details on training and experience.

Are you thinking of a career in technical support? You should have an understanding of the industry beforehand so that you can decide which area you should concentrate on. Once you've decided, then you can tailor your education or training within that specific area.

A technical support career is a good field to be in. There are many different opportunities available that are greatly in demand. Since you will support technology, you'll continue to gain the advanced skills necessary for the future. In addition, the pay structure for technical support positions is very lucrative. These positions require extensive knowledge and experience. As you continue to learn and advance your technical support skills, you will eventually reach the experienced level within a couple of years. To further your career you'll need to continue your professional education via seminars or online and college courses. Attaining industry certifications, such as Microsoft's MCSE will benefit your career. Some certification given by many industry leaders include:

Microsoft's

- MCSE – Microsoft Certified System Engineer
- MCSD – Microsoft Certified Solution Developer
- MCP – Microsoft Certified Professional

Cisco's

- CCNA – Cisco Certified Network Associate
- CCNP – Cisco Certified Network Professional
- CCIE – Cisco Certified Internetwork Expert

As you know, there are many other areas within technical support. Once you've attained and mastered a specific area, you are not obligated to stay there, unless you want to. In many organizations, you have opportunities to transfer into other areas of interest. Given this option, many analysts normally stay with the same company as they continue to advance their technical skills and expertise. Technical support is an exciting industry to be in. And it's evident. You can read about it in many newspapers and magazines. You watch it constantly on television. There are many areas of involvement within technical support and the enormous knowledge you'll receive will give you the freedom to decide on your next interest.

As you contemplate a technical support career, you need to know the different employment options regarding salaries and benefits. There are basically three types of employment options.

- Full-time employee
- Contractor employed by an agency
- Independent contractor

The first option is when a company hires you as a full-time employee. In this case they provide you an industry salary along with benefits such as medical insurance and paid vacation. Many technical analysts prefer this option, because they find comfort in permanent full-time employment and the stability of receiving regular paychecks.

The second option is to be employed by an agency as a contractor. This option allows an analyst a higher rate of pay, although it is not always long-term. The projects are sometimes described as contract or contract-to-permanent. The contract project assignments can run anywhere from a couple of months to a couple of years. The contract-to-permanent assignments can run a few months with the client given the option to hire you on a permanent basis at the end of the project. As an employed contractor you have to negotiate with the agency regarding your pay rate, health benefits and vacations. In most cases the pay rate is generally 30 - 40% more compared to a permanent full-time employee. The health benefits for the most part have to come out of your pocket. Some agencies will cover a small portion (10-20%) of the health benefits premium. As an employed contractor, you must learn to negotiate all aspects of your contract including paid vacation time.

The third option is to be an independent contractor (1099 worker). This is the most difficult option, because you basically go into business for yourself. All expenses will come out of your pocket. However, your pay rate will far exceed the other two previous options, because you don't have to share with an agency. The risks are high when selecting this option, but the rewards are great when you maintain clients. Many analysts do not prefer this option, because of the length of time before you can land a long-term project and the risks. It's takes commitment, determination, and technical proficiency to succeed as an independent contractor.

If you ever decide to become an employed contractor remember, when negotiating with an agency that they will normally offer you 50% of what they expect to bill the client. So if they offer you $25 per hour, they will probably bill the client $50 per hour. I personally do not think this is fair since the contractor does all the work. I understand that the agency has to cover their overhead, but 50% is just too much of a difference. This doesn't mean you have to accept their offer, negotiate further. If you can't negotiate a higher pay rate but the rate they offered you is satisfactory, accept it anyway. While on the project, perform all your responsibilities and exceed all expectations. Then three to six months down the road, approach the agency again and ask for a rate increase. In my experience, I have been able to receive the increase as long as I've performed very well. It is normally best for the agency to grant the increase than to lose the contractor and retrain a new one for the project, costing the client more time and money.

Regardless of which option you choose, it is never good to do average work. Always exceed your performance. Your performance is a major factor in receiving pay increases year after year.

CHAPTER 6

THE FUTURE OF
TECHNICAL SUPPORT

This chapter discusses technical support expectations for the future.

Technical Support will continue to evolve with technology. As demand for state of the art computers and electronics continue, advanced technical support will be needed to support those technologies. That will require more people with advanced technical skills. History dictates that technology will continue to get better. Companies spend millions of dollars every year in research and development to create more advanced technologies. That's because the future of our society depends on technology to discover new boundaries.

In my opinion, there are two areas of technology that will continue to evolve. These areas are:

- Personal Digital Assistants (PDAs) and Handheld Devices
- Network and Wireless Communication

PDAs and Handheld Devices

Today, more and more people are conducting business outside of a main office. They are doing this with the aid of PDAs and handheld devices. These devices allow people to communicate and maintain information to conduct their business and personal activities. There are currently two types of handheld devices: Palm Pilots and Pocket PCs. The Palm Pilots use the Palm Operating System developed by Palm Inc. The Pocket PCs use the Windows CE operating system developed by Microsoft.

What's the difference between these two devices? The main difference is the hardware and functionality. The Palm Pilots manufactured by companies such as 3 Com,

Handspring, Sony, and IBM, allow users the ability to manage their schedule, address book, and to-dos. The Pocket PCs manufactured by such companies as Microsoft, HP and Compaq can also manage the same information. In addition, it has a compact version of the Windows operating system called Windows CE and built-in email capability, Internet Explorer web browser, and other applications such as Pocket Word and Pocket Excel, which are slimmed down versions of its Microsoft Word and Excel counterparts. The Palm also has the ability to send or receive email, but it comes as an added installation.

These portable devices are expandable and are becoming a required accessory for many business people. Many companies continue to make these devices faster and smaller. Therefore, it is only conceivable to assume that the advent of more advanced PDAs will mean added support for the mobile workforce. It is inevitable that support for these mobile individuals will be required wherever they are located. This means that self-service support will be vital to these mobile users. We should see an increase in web-based support provided by more and more companies in the future. The utilization of the internet will continue to rise and become the primary means of communication with the aid of handheld devices.

Network and Wireless Communication

We will also see dramatic changes in the network communications infrastructure, as more data and information requirements will continue to increase the demand for higher bandwidth. With companies doing more business overseas and employees spending less time in the office, the demand for faster remote access to the companies' internal resources will continue to rise. Seamless communications between the network infrastructure, computers, and PDAs will be a priority. The availability of broadband to access office network information will surpass even our expectations, as more people will demand faster and reliable connectivity from their home office.

The utilization of wireless connectivity will become a standard for many companies. That's because it allows users the ability to relocate their laptop computers in the office without having to disconnect from the internal network, allowing users the ability to maintain network connections to printers and servers, therefore, saving time and effort.

Supporting the mobile workforce will be an important topic of discussion for the support organization. They'll need to device a plan for providing the same level of support services as they do onsite. There will be a need for software vendors to discover new ways to improve mobile support for support organizations. Software products such as PC Anywhere will improve and provide ways for support analysts to connect to mobile computers. Hardware vendors will continue to feed the business communities' appetite for handheld devices. People, vendors, and businesses will begin to set standards to allow these technologies to communicate, and the focus will shift from computers to handheld devices.

Through research and development, manufacturers will continue to seek advances in technology, because businesses will demand it. Technical support organizations will continue to make strides to develop better support services because users will need it. As long as there is technology, there will always be a need for the support organization and the support analysts.

About the Author

Jose-Albin D. Afable is president of JDA Systems, Inc. a systems consulting firm based in Chicago, Illinois. He has over 10 years of technical experience in database development, technical support, data analysis, internet related technologies, and customer relationship management. With a decade of technology experience, he founded JDA Systems Inc. to improve small businesses by providing affordable and effective technical solutions.

APPENDIX A

TERMS AND DEFINITIONS

1099 worker

An independent contractor

Bandwidth

The amount and speed of data transferred to and from an
Internet or network connection

Broadband

High-speed Internet access which includes DSL and Cable

Broadband Cable

A type of broadband Internet access, which uses the cable TV
infrastructure to transmit data

Downtime

The times when users are not able to work due to technical problems

DSL (Digital Subscriber Lines)

A type of broadband Internet access, which uses your home's telephone line to transmit data

FAQs (Frequently Asked Questions)

Support materials on websites or in print, which give answers to frequent questions or topics pertaining to products or services

Level 0 Support

Self-service, online, and e-support

Level 1 Support

Helpdesk

Level 2 Support

Field Support

Level 3 Support
The level of support which handles advanced issues, such as server and network connectivity

Network
A group of connected computers that communicate with one another

Network Infrastructure
The network framework and data lines, which allow computers and servers to communicate across a network

Palm OS
The operating system created by Palm, Inc., which is used on all Palm PDAs

Personal Digital Assistants (PDAs)
Small handheld devices or computers

Pocket PC
Handheld devices powered by Windows CE and created by Microsoft

Service Level Agreement
An agreement between the support organization and users, which provide guidelines on how support services will be provided

Survey
A method used to gather feedback

Server
A device used to manage and share files, computer peripherals, and network access

Team Leader
The immediate person (supervisor) in charge of a group of analysts

Troubleshooting
The process of finding the cause and resolution of a computer problem

Windows CE
The operating system developed by Microsoft that power Pocket PC handheld devices

APPENDIX B

LISTING OF COMPUTER HARDWARE AND SOFTWARE MANUFACTURERS

B.1 Computer Hardware Manufacturers

Manufacturer	Website
Acer	www.acer.com
Apple	www.apple.com
Compaq	www.compaq.com
Dell	www.dell.com
Emachines	www.emachines.com
Fujitsu	www.fujitsu-siemens.com
Gateway	www.gateway.com
Handspring	www.handspring.com
HP	www.hp.com
IBM	www.ibm.com

Micron PC	www.micronpc.com
NEC	www.neccomp.com
Palm	www.palm.com
SONY	www.sony.com
Systemax	www.systemax.com
Toshiba	www.toshiba.com

B.2 Computer Software Manufacturers

Adobe

- Acrobat – electronic documents creator
- Illustrator – vector graphics
- Photoshop – image editing

Apple

- Mac OS – operating system
- QuickTime – movie player

IBM

- DB2 - database
- Via Voice – voice recognition

Macromedia

- Dreamweaver – website design
- Fireworks – web graphics
- Flash – rich Internet content and animation

McAfee

- Easy Recovery – file recovery
- Virus Scan – virus scanning and protection

Microsoft

- Encarta - encyclopedia
- FrontPage – website design
- Internet Explorer – web browser
- Microsoft Office – business applications (includes Excel, Word, PowerPoint, Access)
- Money – money management
- Pocket PC – operating system for handheld devices
- Visual Studio – application and database development
- Windows OS – operating system

Symantec

- Ghost – PC cloning and deployment
- Norton Anti-virus – virus scanning and protection
- PC Anywhere – remote control software

APPENDIX C

ONLINE SUPPORT RESOURCES

Askjeeves

www.askjeeves.com

The world's first Internet butler

Avantgo

www.avantgo.com

Provides mobile applications for handheld devices

CNET

www.cnet.com

Provides hardware and software reviews of all industry related products and services

Fast Company Magazine
www.fastcompany.com

An online and print publication dedicated to providing business information for the IT industry

Palm Inc.
www.palm.com

Contains all related hardware and software information for Palm devices including shareware and freeware software for Palm devices

PC Technology Guide
www.pctechguide.com

Provides detailed information about personal computer hardware

PC Magazine
www.pcmag.com

Online computer magazine geared towards personal computers running the Windows operating systems

PCWorld Magazine

www.pcworld.com

Another online computer magazine geared towards personal computers running the Windows operating system

WinDrivers

www.windrivers.com

Resource for Microsoft Windows technical support and drivers

World of Windows Networking

www.wown.com

Provides in-depth information about networking for the Microsoft Windows operating system

APPENDIX D

ONLINE JOB RESOURCES

Brassring
www.brassring.com
Provides listings of current IT jobs

Careerbuilder
www.careerbuilder.com
Provides listings of current job opportunities in many industries

ComputerJobs.com
www.computerjobs.com
Provides substantial technical job listing available by regions

Dice

www.dice.com

An online IT job board listing high tech permanent, contract, and consulting jobs

Headhunter.net

www.headhunter.net

An online job listing service

Hotjobs

www.hotjobs.com

Provides listings of current job opportunities in many industries with numerous job opportunities within the IT industry

Monster

www.monster.com

Provides listings of current job opportunities in many industries with numerous job opportunities within the IT industry

ProSavvy

www.prosavvy.com

Brings clients and consultants together

Realrates

www.realrates.com

Provides salary surveys for computer professionals

Sologig

www.sologig.com

A website geared towards individuals looking for freelance assignments

Techiehunter

www.techiehunter.com

Provides listings of current IT jobs in Chicago and the surrounding suburbs

Vault

www.vault.com

An online career network, which includes career research information and consulting job postings

0-595-22574-8

www.ingramcontent.com/pod-product-compliance
Lightning Source LLC
Chambersburg PA
CBHW051254050326
40689CB00007B/1186